Austrian Special Report

Health, Demography and Climate Change

Summary for Policymakers

(ASR18)

Editors:
Willi Haas, Hanns Moshammer, Raya Muttarak, Olivia Koland

This publication has been published as a pdf version (available at https://verlag.oeaw.ac.at https://epub.oeaw.ac.at/8430-0) and in book form published by the Austrian Academy of Sciences Press.

This report includes the *Summary for Policymakers* of the Austrian Special Report on Health, Demography and Climate Change (ASR18 - full volume in German language available at https://epub.oeaw.ac.at/8427-0). The Special Report contains references to online supplements, which provide further texts on selected contents of the report (available at https://epub.oeaw.ac.at/8464-5).

Citation of the "Summary for Policymakers": Haas, W., Moshammer, H., Muttarak, R., Balas, M., Ekmekcioglu, C., Formayer, H., Kromp-Kolb, H., Matulla, C., Nowak, P., Schmid, D., Striessnig, E., Weisz, U., Allerberger, F., Auer, I., Bachner, F., Baumann-Stanzer, K., Bobek, J., Fent, T., Frankovic, I., Gepp, C., Groß, R., Haas, S., Hammerl, C., Hanika, A., Hirtl, M., Hoffmann, R., Koland, O., Offenthaler, I., Piringer, M., Ressl, H., Richter, L., Scheifinger, H., Schlatzer, M., Schlögl, M., Schulz, K., Schöner, W., Simic, S., Wallner, P., Widhalm, T., Lemmerer, K. (2018). Austrian Special Report (ASR18) Health, Demography and Climate Change. Austrian Panel on Climate Change (APCC), Vienna, Austria, 26 pages.

This publication was created under the umbrella of the Austrian Panel on Climate Change (APCC, www.apcc.ac.at) and follows its quality standards. As a result, the Special Report has undergone a multi-stage review process with the intermediates *Zero Order Draft, First Order Draft* and *Second Order Draft.* In the last international review step of the *Final Draft,* the incorporation of review comments was reviewed by review editors.

Austrian Panel on Climate Change (APCC):
Helmut Haberl, Sabine Fuss, Martina Schuster, Sonja Spiegel, Rainer Sauerborn

Co-Chairs and Project Lead:
Willi Haas, Hanns Moshammer, Raya Muttarak, Olivia Koland

Review-Management:
Climate Change Center Austria (CCCA)

Cover:
Alexander Neubauer

Translation of the "Summary for Policymakers" into English:
Senta Movssissian-Knor, translate that

 The APCC Report was prepared with the participation of scientists and experts from the following institutions: Federal Ministry for the Environment, Nature Conservation and Nuclear Safety (D), Climate Change Center Austria (CCCA), Health Austria GmbH, Helmholtz Center for Environmental Research, Medical University of Vienna, Austrian Agency for Health and Food Security, Austrian Academy of Sciences, Potsdam Institute for Climate Impact Research, Statistics Austria, Graz University of Technology, Federal Environmental Agency, University of Augsburg, University of Natural Resources and Life Sciences Vienna, University of Vienna, University of Nottingham, Wegener Center for Climate and Global Change of the University Graz, Vienna University of Economics and Business, Wittgenstein Center for Demography and Global Human Capital, World Health Organization and Central Institute for Meteorology and Geodynamics.

We sincerely thank the authors for their dedicated work, which went far beyond the scope of the funded project (see author list on the next page). Many thanks to the participants of the two stakeholder workshops, who provided important input for the focus of the report. We also thank Zsófi Schmitz and Julia Kolar for the professional review management and Thomas Fent for his support in the final process of the implementation of the book project.

The Special Report ASR18 was funded by the Climate and Energy Fund as part of its funding program ACRP with around 300,000 euros.

Archivability tested according to the requirements of DIN 6738, lifespan class – LDK 24-85.

Austrian Academy of Sciences Press, Vienna

ISBN 978-3-7001-8430-0
https://verlag.oeaw.ac.at
https://epub.oeaw.ac.at/8430-0

ISBN (full volume in German) 978-3-7001-8427-0
https://verlag.oeaw.ac.at
https://epub.oeaw.ac.at/8427-0

Layout: Berger Crossmedia, Wien
Print: Gugler*print, Melk/Donau, made in the EU

Except Cover and Binding

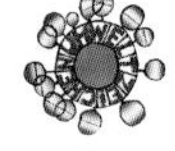

Printed according to criteria documents of the austrian Eco-Label „printed products". gugler*print, Melk, UWZ-Nr. 609, www.gugler.at

Summary for Policymakers

ASR18

Authors:

Co-Chairs
Willi Haas, Hanns Moshammer, Raya Muttarak

Coordinating Lead Authors (CLAs)
Maria Balas, Cem Ekmekcioglu, Herbert Formayer, Helga Kromp-Kolb, Christoph Matulla, Peter Nowak, Daniela Schmid, Erich Striessnig, Ulli Weisz

Lead Authors (LAs)
Franz Allerberger, Inge Auer, Florian Bachner, Maria Balas, Kathrin Baumann-Stanzer, Julia Bobek, Thomas Fent, Herbert Formayer, Ivan Frankovic, Christian Gepp, Robert Groß, Sabine Haas, Christa Hammerl, Alexander Hanika, Marcus Hirtl, Roman Hoffmann, Olivia Koland, Helga Kromp-Kolb, Peter Nowak, Ivo Offenthaler, Martin Piringer, Hans Ressl, Lukas Richter, Helfried Scheifinger, Martin Schlatzer, Matthias Schlögl, Karsten Schulz, Wolfgang Schöner, Stana Simic, Peter Wallner, Theresia Widhalm

Contributing Authors (CAs)
Franz Allerberger, Dennis Becker, Michael Bürkner, Alexander Dietl, Mailin Gaupp-Berghausen, Robert Griebler, Astrid Gühnemann, Willi Haas, Hans-Peter Hutter, Nina Knittel, Kathrin Lemmerer, Henriette Löffler-Stastka, Carola Lütgendorf-Caucig, Gordana Maric, Hanns Moshammer, Christian Pollhamer, Manfred Radlherr, David Raml, Elisabeth Raser, Kathrin Raunig, Ulrike Schauer, Karsten Schulz, Thomas Thaler, Peter Wallner, Julia Walochnik, Sandra Wegener, Theresia Widhalm, Maja Zuvela-Aloise

Junior Scientists
Theresia Widhalm, Kathrin Lemmerer

Review Editors
Jobst Augustin, Dieter Gerten, Jutta Litvinovitch, Bettina Menne, Revati Phalkey, Patrick Sakdapolrak, Reimund Schwarze, Sebastian Wagner

Austrian Panel on Climate Change (APCC)
Helmut Haberl, Sabine Fuss, Martina Schuster, Sonja Spiegel, Rainer Sauerborn

Project Lead
Willi Haas und Olivia Koland

Table of contents

Bundespräsident
Alexander Van der Bellen

As early as in the mid 1980s it became clear
that the climate crisis would not come as a slow, linear development,
but would rather turn into a global challenge at a rapid pace.
In the meantime, the effects are being clearly seen and felt across the globe.
The primary objective is still to reduce greenhouse gas emissions.
But we already have to protect ourselves against the health effects of climate change.
Great heat, extreme drought, heavy precipitation, hurricanes, floods, and, of course,
also the indirect consequences of these phenomena are affecting all of us, humankind as a whole.
In addition to scientific findings,
an adequate response to the climate crisis also requires political action:
in international politics, European politics, national, regional and local politics.
I believe no one has any illusions about how difficult it is
to take political action that is geared towards the urgently needed transition to a different,
ecologically oriented, climate-friendly and health-promoting global society.
This Austrian Special Report now presents a coordinated scientific assessment that provides a basis
for far-reaching political decisions.
I wish for this report, and for the sake of all of us, that political action will follow.
I would like to thank all the authors
in their various roles, the review management, reviewers and international review editors,
management and stakeholders as well as the APCC members
for their dedicated, future-oriented contributions.

Federal Minister for Sustainability and Tourism

Climate change has arrived in our midst; its effects are already being clearly felt.

Glaciers are melting, the sea level is rising, heat waves, droughts and other extreme weather events are on the increase across the globe. The consequences already have a massive impact on people's lives and their economic activities. The effects on human health are evident and are the subject of this report. Climate protection also entails doing something beneficial for our own health.

The Austrian Federal Government attaches great importance to climate protection. With #mission2030, the Austrian Climate and Energy Strategy, we have laid the foundation for numerous measures and strategies that are now being implemented and represent steps in the right direction. Through the klimaaktiv participatory climate protection initiative we support businesses, households and municipalities in putting effective climate protection measures into practice.

Climate protection is a huge challenge that will remain an issue for generations to come. Effective climate protection is based on the two pillars of energy efficiency and renewable energies. It affects all areas of life - how we live, work, dwell and move. In the long run, the costs for the protection of our climate will be significantly lower than those caused by unchecked global warming. Thus, it is all the more important for us to systematically consider the consequences of climate change in all relevant planning and decision-making processes now.

For many years already, Austria has intensified its efforts in dealing with the question of how to best respond to climate change. In this study, commissioned by the Austrian Climate and Energy Fund, profound facts have been produced. What we need now are specific solutions to be prepared for the future. We must resolutely implement Austria's strategy for adapting to climate change, jointly supported by the federal and state governments, and prepare ourselves for future challenges in the best way possible.

Key Statements

Climate change and its health effects

- The health consequences of climate change are already being felt today and are to be considered an increasing threat to health.
- The most severe and far-reaching health effects are to be expected as a consequence of heat.
- Climate change leads to increased health effects associated with pollen (allergies), precipitation, storms and mosquitoes (infectious diseases).
- Demographic change (e.g. aging) increase the population's vulnerability and thus intensify climate-induced effects on health.

Addressing the health effects of climate change and reducing vulnerability

- **Heat:** Heat warning systems complemented by action-oriented information for persons who are hard to reach can become effective at short notice; urban development measures have a long-term effect.
- **Allergens:** Fighting highly allergenic plants reduces health effects and therapy costs.
- **Extreme precipitation, drought, storms:** Integrated event documentation for more targeted measures, strengthening self-provision and involvement of diverse groups in the preparation of contingency plans can help lessen the impact.
- **Infectious diseases:** Promoting capacities in early detection among the population and health staff for preventive purposes; targeted control of invasive species in order not to endanger other species.
- Growing **health inequality** of vulnerable groups induced by climate change can be avoided by strengthening health literacy.
- Promoting **climate-specific health literacy** of health staff as well as enhancing the **quality of dialogue** with patients for the individual handling of climate change and developing healthier and more sustainable lifestyles (diet, exercise).
- Systematically promoting children's/young persons' **education and training** to develop an understanding of climate and health-relevant issues, allowing them to act accordingly.

Leveraging opportunities for climate and health

- **Diet:** The positive implications of a reduction in excessive meat consumption in particular are considerable in terms of climate protection and health, with comprehensive sets of measures, including price signals, showing positive effects.
- **Mobility:** A shift to more active mobility and public transport, in particular in cities, reduces air and noise pollution and leads to healthy movement; reduction of climate-relevant air traffic also diminishes adverse health effects.
- **Housing:** The high percentage of newly built single-family and duplex houses is to be challenged as it uses a lot of space, materials and energy, and making apartment buildings attractive as an alternative to a house in a green area requires funding; pushing health-enhancing and climate-friendly urban planning; thermal renovation reduces the heat stress during the summer half-year.
- **Health sector:** The climate-relevance of this sector makes a specific climate strategy necessary; pharmaceutical products are responsible for a major share of the carbon footprint; avoiding unnecessary diagnostics and therapies reduces greenhouse gas emissions, risks for patients and health-related costs.

Initiating transformation at the intersection of climate and health

- **Cross-policy collaboration in the field of climate and health policies** represents an appealing opportunity to simultaneously implement Austria's Health Targets, the Paris Climate Agreement and the United Nations Sustainability Goals.
- **Harnessing the scientific potential for transformation:**
  - Innovative methods in science, like, for instance, transdisciplinary approaches, can trigger learning processes and make accepted problem-solving more likely.
  - Research in medicine and agriculture has to become more transparent (funding and methods); issues such as the reduction of over-medication and multiple diagnoses or the health-related assessment of organic food require independent funding.
  - Learning from health promoting and climate-friendly everyday practices of local initiatives, like, for instance, eco-villages, slow food, slow city movements and transition towns.
  - Transformation research and research-oriented teaching accelerate transformative development paths and encourage new interdisciplinary solutions.

1 Challenge and Focus

The effects of climate change on health are already being felt today. Based on current projections of future climate trends, it is to be expected that the world population will have to face unacceptably high health risks. This is obvious from both the most recent report of the IPCC as well as more recent papers published by leading experts. In Austria, the effects of climate change can already be observed and are to be considered a growing threat to health, which will be further intensified by demographic change.

The assessment presented here summarizes the state of scientific knowledge on the topics of "climate-health-demography". The assessment starts out with the issues of climate, population, economy and health care as interacting determinants of health (Fig. 1). In this context, climate change either has direct effects on health, like, for instance, during heat waves, or indirectly through changing natural systems, such as through an increased release of allergens or more favorable living conditions for disease-transmitting organisms. The extent to which climate change will eventually affect health, however, can be estimated only in combination with population dynamics, economic development and health care. A higher proportion of elderly people or of the chronically ill, poorer health care or also an increasing number of people with lower income lead to a higher level of vulnerability of society to climate change.

There are various options for action available to the government as well as businesses and individuals. If the goal is to create a largely climate-neutral society, it seems necessary to make use of a multitude of these options for action. In addition to individual climate protection measures, however, a more extensive transformation toward a climate-friendly society is necessary that addresses the underlying causes of climate change. This approach often entails an added health benefit of climate protection measures (co-benefits). At the same time, in light of progressing climate change, measures to adapt to climate change have to be taken to minimize the adverse consequences for health.

In order to make a reliable assessment of these complex causal relationships that are also relevant to Austria, a transparent process, comprehensive in terms of content and taking account of an interdisciplinary balance, was implemented for the preparation of an Austrian Special Report following the style of the Austrian Assessment Report Climate Change (AAR14) and the reports of the Intergovernmental Panel on Climate Change (IPCC). More than 60 scientists made contributions as authors and another 30 as reviewers to provide a basis for decision-making in the fields of science, administration and politics, facilitating efficient and responsible action.

The key finding of this work of one and a half years is that a well-coordinated climate and health policy can be a powerful stimulus for transformation toward a climate-compatible society that promises a high level of acceptance owing to its potential to bring about better health and a higher quality of life for all.

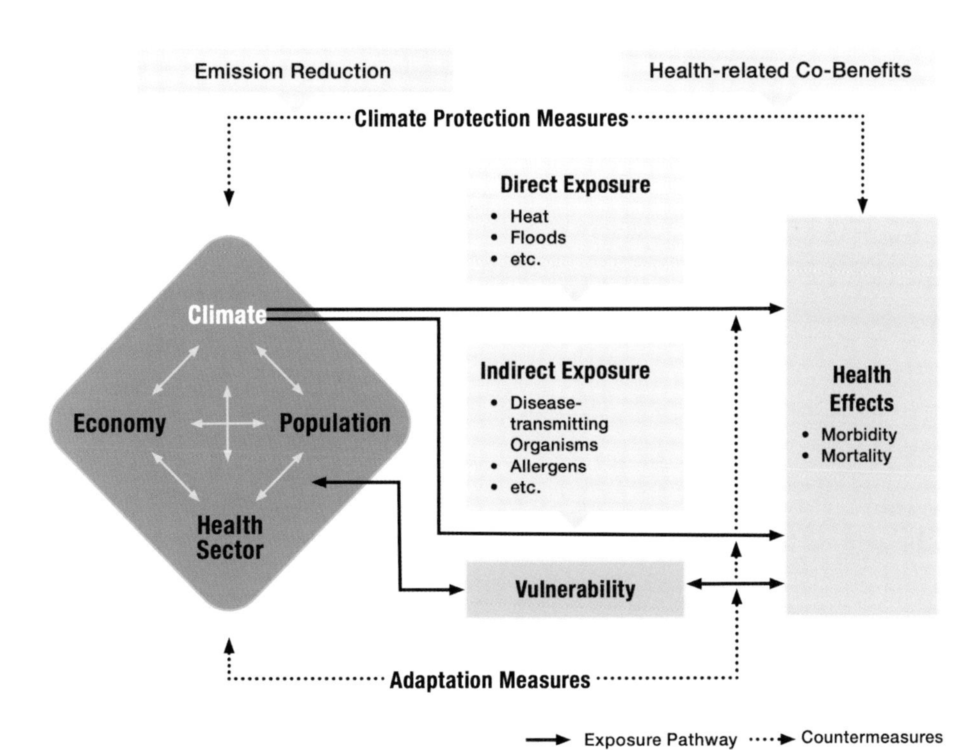

Fig. 1: Dynamic model how changes in health determinants affect health.

2 Health-relevant Changes in the Climate

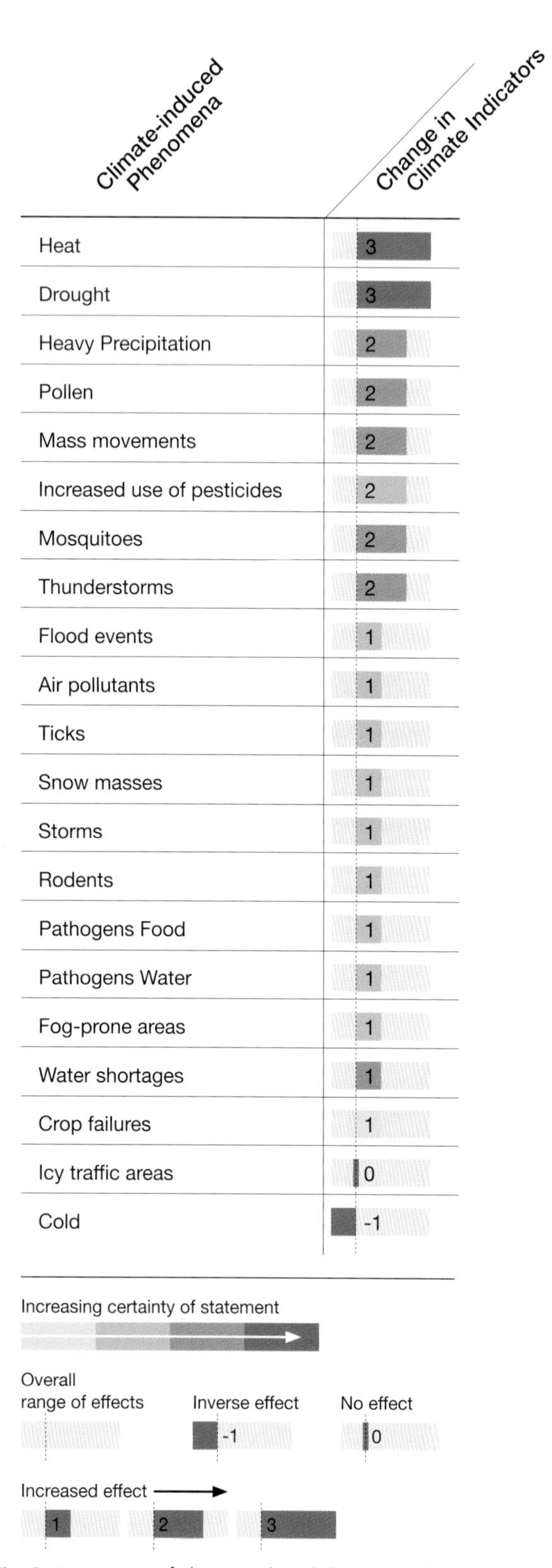

Fig. 2: Assessment of climate-induced changes in health-relevant phenomena with a time horizon until 2050 (3 = major adverse changes)

To better understand the health-related relevance of climate change in a first step, those climatic phenomena which have an effect on health were identified. For this purpose, climatologists made an estimation of the climatic changes to be expected by 2050, initially without taking into account how many persons will be affected and to what extent. In doing so, the uncertainty of the statement with regard to climate change was also taken into consideration (Fig. 2).

The most severe changes posing the greatest threat to health are to be expected during heat waves, both because of the steady rise in temperature during the summer half-year, the number of hot days, the duration of the heat events and a less pronounced drop in temperature at night. Drought also falls into the category of major changes. In this context, however, it turns out that owing to its sound food supply, Austria will presumably see only minor health effects. Both during heat and drought the climatological statements show high certainties. With regard to severity and certainty of statement, extreme precipitations are rated slightly lower. Based on reliable statement, an increasing incidence of allergies is rated as highly relevant, with prolonged seasons, an increased occurrence of already indigenous allergenic plants and migration of new allergenic plant and animal species expected with medium certainty. In all areas mentioned, climate change leads to an aggravation of adverse health effects – except for the cold. The number of cold days is declining, the duration of cold periods is decreasing and it is highly certain that average temperatures will rise during the winter half-year. Thus, it can be inferred that the number of cold-related conditions and/or cold-related deaths will decrease, which, however, will not outweigh the adverse effects of an increased number of heat waves. In addition, there is a risk not described here, i.e. that the melting of the arctic ice and a resulting slowdown of the Gulf Stream may lead to longer and colder winters with an increased number of cold-related deaths also in Austria.

For all climatic phenomena outlined here, the consequences may differ strongly from region to region and may be different in rural areas and in metropolitan areas.

3 Priorities regarding Climate-induced Health Effects

To allow a better classification of the urgency regarding the various health-relevant developments, based on their state of knowledge, 20 key experts of the Assessment Report evaluated them based on two groups of criteria:

- Affected persons: proportion of affected persons among the population taking into account socio-economically disadvantaged groups and vulnerable persons, such as infants, elderly people and persons with pre-existing illnesses.
- Health effects: mortality, physical and mental morbidity

According to this assessment, the highest priority is to be given to situations where the combined effect of both criteria groups occurs, i.e. when a relatively high percentage of the population has to expect serious health effects. The differing estimates within the individual criteria explain the grading of the ascribed priority levels. In addition, possible options for action at individual and government level (the latter also includes the health care system) were assessed. This expert assessment is to be seen as cross-topic and thus integrative guidance – it cannot replace a strictly scientific analysis.

The assessments showed a clear categorization into three priority levels according to which the individual issues should be addressed (Fig. 3): Heat is at the top of the table with the highest priority, followed by pollen and air pollutants together with extreme events such as heavy precipitation, drought, flood events, mudslides and landslides. Little significance, on the other hand, is attributed to cold-related events, shortage of water or food and pathogens in water and food. The high priority assigned to the group of "air pollutants" is remarkable, although uncertainties as to further developments are high. As the collective term comprises both ozone (upward tendency) and particular matter concentrations (downward tendency) the findings are hard to interpret. The events particularly affecting economically disadvantaged persons as well as elderly and ill persons mainly fall into the highest priority category. Any adverse health effects of crop failures are less likely in Austria on account of the favorable supply situation – if necessary, supported by imports.

Figure 3 clearly shows that both at individual and government level options for action are identified – generally more at government level. No differentiation was made between these as to their nature, i.e. preventive measures, crisis interventions and follow-up measures are included. Not all measures are part of the health care system, as the example of an imminent increased use of pesticides in farming shows. Only in one single case is the individual person thought capable of dealing with more options for action than the government – namely in connection with icy traffic areas.

In the statements below, special attention is paid to the fact that many of the measures that are important in terms of climate protection have positive "side effects" (co-benefits). This is especially true for health, which is why the measures are warranted even in the absence of any climate effect.

Priority Level	Climate-induced Phenomena	Trigger Events and/or Potential Health Effects	Change in Climate Indicators	Proportion of Affected Population Groups	Effect on Socially Weak Groups	Effect on Elderly and Ill Persons	Level of Health Effect (Morbidity/Mortality)	Individual Options for Action	Government Options for Action
3	Heat	Steady increase and more, longer, hotter heat waves, less pronounced drop in overnight temperature	3	3	++	+++	3	2	2
2	Pollen	Extended seasons and more allergenic neobiota	2	2	+	+	2	1	1
2	Air pollutants	Climate-induced increased effect of ozone, decrease in particulate matter	1	2	+	++	2	1	2
2	Heavy precipitation	More frequent and intense	2	1,5	+	+	2	1	2
2	Drought	Water and food shortage	3	1	++	++	2	1	2
2	Flood events	More frequent and intense	1	1,5	+	+	2	2	2
2	Mass movements	Mudslides and landslides	2	1	+	+	2	1	2
1	Increased use of pesticides	Due to increased occurrence of vermin	2	2	+	+	1	1	3
1	Mosquitoes	Malaria	2	1	+	+	2	1	2
1	Thunderstorms	Increased and fiercer	2	1	+	+	2	1	1
1	Ticks	Surge in ESME/TBE, Lyme disease	1	1		+	2	2	2
1	Snow masses	Increasing events	1	1		+	2	1	1
1	Storms	Increased and stronger whirlwinds and tornadoes	1	1	+	+	2	1	1
1	Rodents	Leptospirosis, HFRS, tularemia	1	1		+	2	1	1
1	Pathogens Food	Campylobacter, salmonella, *E. coli* & vibrio infections, mycotoxins	1	1	+	+	1	2	2
1	Pathogens Water	*Giardia lamblia*, *E. coli*, vibrio and cryptosporidium infections	1	1			1	1	2
1	Fog-prone areas	Risk of accidents	1	1			1	1	1
1	Crop failures	Food shortages	1	1	+	+	1	1	2
1	Water shortages	Less accumulation of ground water	1	1	++	+	0	0	2
0	Icy traffic areas	Risk of accidents	0	0		+	1	2	1
0	Cold	Frostbite, strain on the immune system	-1	-1	++	+	2	2	2

Fig. 3: Assessment of impacts of climate-induced phenomena on health with a time horizon until 2050 (3 = major adverse changes) for a share of the affected population as well as the extent of health effects sorted by urgency categories (3 = highest need for action)

4 Mitigating Health Effects

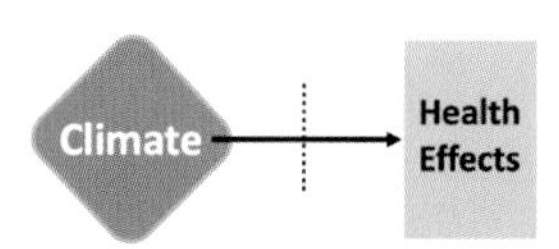

This section summarizes the developments and effects of the most pressing climate-induced health effects as well as options for action for their prevention with regard to Austria. Moreover, it addresses basic strategies for dealing with increased vulnerability due to demographic dynamics as well as ways and means of reducing vulnerability.

4.1 Addressing Climate-induced Effects

Heat

Climate: By the middle of this century, it is to be expected that the number of hot days, i.e. days during a heat episode (periods with daily maximums of 30 °C and above), will double; by the end of the century, if no sufficient climate protection measures are taken, there may be a tenfold increase in the number of hot days. Less of a drop in evening temperatures also adds to the problem; the number of nights during which temperatures do not fall below 17 °C has gone up by 50 percent in Vienna (comparison of 1960–1991 and 1981–2010) (high agreement, robust evidence).

Health: Assuming that no further adaptation is made and climate change will be moderate, 400 heat-related deaths per year can be expected in Austria in 2030; by mid-century, this number could rise to 1,000 cases per year, with the major share occurring in cities (according to more recent climate projections, 2030 could see even higher figures). Elderly people and persons with pre-existing illnesses are particularly vulnerable, economically weaker persons and/or migrants are often more strongly affected due to their housing situation (densely built-up areas, few green areas) (high agreement, medium evidence).

Options for action: Swift implementation of urban development measures to alleviate the problem of urban heat islands, planting trees, enhanced air circulation, reduction of the thermal load of heat-generating sources, facilitating cooling down at night, reduction of air pollutants and noise pollution allowing rooms to be aired at night may lead to significant improvements in the long run and may help avoid energy-consuming air conditioning that is potentially harmful to the climate. In the short term, an evaluation of the heat warning systems may be advisable, by focusing in particular on action-oriented information of persons who are hard to reach (e.g. elderly people without Internet access or persons with language barriers) in cities (high agreement, robust evidence).

Allergens

Climate: Climate change in combination with globalized trade and travel as well as land use change lead to the spread of plant and animal species that were previously not indigenous, but that have an impact on health. A significant increase in the pollen count caused by ragweed (*Ambrosia artemisiifolia*) is expected, which is enhanced by increased air humidity as well as the "fertilizing effect" of CO_2 and nitrogen oxides. The German Assessment Report expects the emergence of another six new plant species with clear potential for being harmful to health. In urban areas in particular, the concentration of pollen in the air has surged (high agreement, robust evidence).

Health: As a consequence, the number of respiratory diseases (hay-fever, asthma, COPD) is rising. Increased health effects can be expected particularly in urban areas in combination with air pollutants (ozone, nitrogen oxides, particulate matter, etc.) as they lead to an increased allergenic aggressiveness of pollen. Even today, approximately 1.75 million people in Austria suffer from allergic diseases. The frequency and severity of allergies will increase. According to estimates, 50 percent of all Europeans will be affected in 10 years' time (high agreement, medium evidence).

Options for action: Planned nationwide monitoring can alleviate adverse effects through targeted information. By consistently fighting highly allergenic plants (e.g. mowing or weeding before seed formation in *Ambrosia*), health effects can be avoided and considerable therapy costs can eventually be saved. This is, for instance, documented in analyses of the health effects of the spread of *Ambrosia* in Austria and Bavaria (high agreement, medium evidence). The incorporation of control measures into law, and the involvement of key stakeholders, can contribute significantly to the reduction of the health effects caused by *Ambrosia* in Austria.

Extreme precipitation, drought, storms

Climate: For physical reasons, more intense and abundant rainfall, longer periods of draught and heavier storms can be expected in the wake of climate change (medium agreement, medium evidence). Even today, costs for damage caused by extreme events are considerable in Austria, and soaring.

Health: Extreme weather events make good headlines, but the number of people exposed to them is – disregarding extreme temperature events – relatively small, thus the immediate health effects of extreme weather phenomena in Austria are relatively low (high agreement, medium evidence). Nevertheless, extreme events can cause immediate health effects, such as injuries or deaths and, in particular in the case of existence-threatening material damage, post-traumatic stress disorders. Indirectly, poor water quality following floods can trigger bacterial infections. Extreme weather events occurring in other countries may lead to (climate-induced) migration;

presently, however, this is not considered a serious threat to the Austrian population's health thanks to the high standard of Austria's health care system.

Options for action: An integrated event documentation (merging of the reliable records on the initial situation, causes, measures, effects) can facilitate the analysis and preparation of bespoke measures (high agreement, medium evidence). Damage and health effects can be reduced further by strengthening self-provision and by well-coordinated collaboration in the risk management of public and private stakeholders. These can be supported by including them in school curricula, targeted information, advisory services and incentives for preventative disaster management, like, for instance, technical and financial support as well as reduced insurance premiums for well-prepared households. The involvement of a good mix of diverse groups can be beneficial in the preparation of effective disaster management plans, in particular at municipal level, as both their needs are taken into account and their potential is used in dealing efficiently with disasters (medium agreement, medium evidence).

Infectious diseases

Climate: Climate change (in particular global warming) will have an effect on the occurrence of mosquitoes as vectors of diseases, as subtropical and tropical mosquito species introduced to Austria (especially of the *Aedes* species: tiger mosquito, Asian bush mosquito, etc.) will find better survival conditions here in future. Thus, they will inhabit more extended areas, particularly spreading across the northern and altitudinal limits. Some of our indigenous mosquito species may also transmit pathogens of infectious diseases previously rarely seen in Austria, such as the West Nile virus or the Usutu virus. Moreover, an increased distribution of sand flies and wood ticks (*Dermacentor* ticks) as potential vectors of several infectious diseases (leishmaniasis, FSME virus, Crimean-Congo hemorrhagic fever virus, Rickettsia, Babesia, etc.) could be observed.

Health: The occurrence of infectious diseases is determined by complex interrelations, ranging from globalized traffic, people's temperature-dependent behavior and local weather factors (e.g. humidity) to the survival rate of infectious agents – depending on the water temperature (high agreement, medium evidence). The concrete relationships, however, have not yet been sufficiently studied to allow conclusive statements to be made. Furthermore, if global warming progresses, diseases related to food may occur (e.g. campylobacter and salmonella infections, contamination with mycotoxins); the high national food production standards, however, – in particular well-functioning cold chains – do not give us reason to expect any significant implications for the incidence of these diseases in Austria in the near future (high agreement, medium evidence).

Options for action: A pivotal factor in combating infectious diseases in time is early detection. This can be improved, on the one hand, by promoting the relevant health literacy among the population, and, on the other hand, by further developing professional competency in health professions, particularly in primary care. In doing so, the climate-induced infectious diseases that are essentially highly treatable can be detected quickly, despite their previously rare occurrence (high agreement, robust evidence). In this context, the re-orientation of the public health service as laid down in *Zielsteuerung-Gesundheit 2017* (Target-based Health Governance) can have a supporting effect (establishing nationwide pools of experts in new infectious diseases). To achieve the best possible control measures, the evaluation and exchange of knowledge at international level are of vital importance. In addition, special attention has to be paid to the targeted fight against dangerous species in order not to deprive amphibians and other animals of their basic food resources by exterminating non-hazardous insects (e.g. non-biting midges) (high agreement, medium evidence). In the field of food, adapted food monitoring for climate change-related monitoring and – if necessary – adaptation of the guidelines regarding best agricultural and hygienic practices can contribute to health protection. It should be pointed out that the use of disinfectants can have negative effects on the environment and humans and is, in most cases, especially in households, absolutely unnecessary. There is a need for research with regard to possible extensions of propagation areas of potential vectors. A review of food monitoring and – if required – its adaptation in Austria by AGES can contribute further to food safety.

4.2 Reducing Vulnerability

Mitigating the Aggravating Effect of Demographic Change on Climate-related Health Effects

Development dynamics: Austria's population is growing mainly in the urban regions. On average, the aging of the population is characterized by a shrinking proportion of people of working age and a consistent proportion of children and adolescents. Aging is mitigated by the migration of young adults. The figures for the outskirts show a decline in population for educational and employment reasons, while at the same time the population is increasingly aging. In the long run, an annual migratory balance for Austria of approximately 27,000 additional people (period 2036–2040) is to be expected. While exact figures are uncertain, net migration is likely to increase (high agreement, robust evidence). It is to be assumed that the incidence rate of chronic diseases like, for instance, dementia, respiratory diseases, cardiovascular diseases and malignant tumors (malignomas), including all health implications following in their wake, will go up. It is quite remarkable that the occurrence of more than half of all mental disease cases can be witnessed among the age group of over 60-year olds.

Relation to climate: Due to the high proportion of cardiovascular diseases, diabetes and mental diseases in people over 60 years of age, older population groups are especially vulnerable to the impact of climate change, in particular heat. In addition, future, more frequent extreme weather events are likely to lead to increased mental stress for elderly people. Persons with only few resources at their disposal are particularly susceptible to the negative effects of climate change. These include, for example, poor education and insufficient financial means, structural, legal and cultural barriers, limited access to health infrastructure or unfavorable housing conditions. Refugees, in particular, are highly vulnerable due to the deprivations they have experienced and as a consequence of the physical and mental stresses and strains related thereto. The health risk of transmitting imported diseases, however, is extremely low, even in cases of close contact.

Options for action: Targeted measures to strengthen health literacy among particularly vulnerable target groups, such as elderly people and persons with a migration background, can combat the climate-induced aggravation of inequality. To this end, multiculturalism in health facilities can be used for multilingual communication and transcultural medicine and care through targeted diversity management (high agreement, medium evidence). In particular, target group-specific prevention, promotion of health and treatment, as well as further development of the living conditions of vulnerable groups, can mitigate any further exacerbation of unequally distributed disease burdens – especially with regard to heat and mental health, according to the "Health (and Climate) in all Policies" approach. Such measures can be enhanced by accompanying and supplementary research.

Counteracting Aggravated Health Inequalities Induced by Climate Change

Development dynamics: 14 percent of the Austrian population are at risk from poverty and marginalization. Large families, lone-parent households, migrants, women of retirement age, the unemployed, unskilled workers and people with low education levels suffer from a significantly increased risk of falling into poverty. Even today, socio-economic inequality is contributing to health imbalances: In Austria, persons with no more than compulsory schooling show a life expectancy 6.2 years shorter than that of university graduates (high agreement, robust evidence).

Relation to climate: These health inequalities are in many ways fueled by climate-related changes (high agreement, medium evidence). Exposed workplace and housing conditions have an additional exacerbating impact (such as, for instance, heavy outdoor work on construction sites or in agriculture, lack of urban green spaces near people's homes, high exposure to noise pollution in living quarters). In the past, heat waves and natural disasters have already affected disadvantaged groups more directly. Add in further vulnerability factors (such as old age), and the situation will deteriorate (high agreement, medium evidence). The heat wave that struck Vienna in 2003, for instance, led to a significant increase in the number of fatalities in low-income districts in particular. As yet, the impacts of climate change on health inequalities have hardly left their mark on "Health in all Policies" approaches. Globally, unequal exposure to health risks induced by climate change has been identified as a key factor. As a consequence, the UN Sustainable Development Goals (SDGs) address the correlation between socio-economic status, health and climate. By contrast, strategic and political discussions in Austria tend to overlook it.

Options for action: Building on the measures of the Austrian Health Target 2 "Fair and Equal Opportunities in Health", in particular with regard to poverty alleviation, the aggravating factors of climate change can be cushioned by way of targeted support measures in the fields of life and work environments. By setting up a coordination and exchange platform mirroring a "community of practice", it is possible to support the hands-on approach when implementing these measures (medium agreement, low evidence). During the implementation of the sustainability goals at public administration, governmental and other societal levels (economy, civil society) in Austria, it is possible to further deepen the coordination of cross-policy cooperation to promote fair and equal opportunities. Interdisciplinary research projects on health-related equal opportunities in the light of climate change play a pivotal role (high agreement, low evidence) and may provide insights as to which targeted measures to take in order to bring back into balance health-related inequalities of particularly disadvantaged groups and very strongly affected regions.

Developing Climate-related Health Literacy for the Purpose of Reducing the Impact of Climate Change

Development dynamics: Acquiring a high level of health literacy enables any individual to better understand physical and mental health issues and to decide wisely in health matters. Low levels of health literacy will lead to low levels of treatment adherence, delayed diagnoses, poor self-management skills and an increased risk of developing chronic diseases. Consequently, poor health literacy results in high health care costs. According to an international survey, more than 50 percent of the Austrian participants have developed only inadequate or problematic health literacy skills. These figures were as high as 75 percent for people of poor health, with little money or older than 76 years of age. The survey shows that mostly this is not due to the individuals' cognitive skills, but rather to various aspects of the health care system (medium agreement, medium evidence). The health reform entitled *Zielsteuerung Gesundheit* (Target-based Health Governance) and the Austrian Health Targets took note of this predicament and set out operative targets. What they fail to acknowledge, however, is the role climate change plays in terms of people's health.

Relation to climate: Disadvantaged groups fall victim to climate change much more often and, in addition, they often show lower levels of health literacy, while at the same time, the information material available does not reach them quite as readily (high agreement, medium evidence). The action plan of the Austrian adaptation strategy already points to existing education and information projects, especially with regard to health topics. Also, from a climate protection perspective, a healthier diet and health-enhancing physical activity in everyday life can contribute to lowering greenhouse gas emissions.

Options for action: Improving climate-related health literacy may result in a reduction of the impact climate change has on the health of particularly vulnerable groups and even improve people's health. What is required to this end is for all competent health and climate-related authorities both at federal and provincial levels to join forces in intersectoral cooperation (high agreement, medium evidence) following the maxim: the more target group-oriented the measures, the better the effects to be achieved. Thus, it would be of vital importance to systematically teach health care professionals about climate-specific health care issues as part of their education and training, because it is they who can address individual health concerns and give appropriate individualized advice and assistance and who can initiate lifestyle improvements in people's living environments. Key topics of relevance to the climate in this respect are: heat, also in combination with air and noise pollution, allergies, (newly emerging) infectious diseases as well as diet, mobility and local recreation. All this will help to create a widespread dissemination of climate-related health literacy, especially through personal talks and counseling on climate-friendly health behavior (e.g. active mobility and healthy diet). Health professionals, most notably physicians, are called upon to act as "personal health advocates". Measures meant to improve the quality of dialogue in patient care (education and training) can be extended to also include the climate change aspect. By promoting targeted educational measures in schools (curricula and teaching practice), children and adolescents can be taught to adopt a climate- and health-conscious lifestyle.

5 Leveraging Opportunities for Climate and Health

Apart from identifying impending threats, measures can be taken in areas suited to generating benefits for both climate and health. By fine-tuning political instruments, actions beneficial to climate and health can be made more appealing and detrimental actions can be made less worthwhile, which could lead to changes within otherwise problematic areas as well.

5.1 Diets

Need for action: From a health perspective, a dietary change would be in order, with a particular focus on reducing excessive meat consumption for climate and health reasons. In Austria, meat consumption significantly exceeds the healthy levels recommended by the Austrian food pyramid, with amounts for adult males, for example, tripling the recommended levels, while the share of cereals, fruit and vegetables is too low (high agreement, robust evidence). Like many other countries, Austria is experiencing an increase in nutritional diseases. The consumption of animal products significantly increases the risk of developing diabetes mellitus type 2, high blood pressure and cardiovascular diseases. The implementation of the United Nations' Sustainable Development Goals (SDGs) also requires changes in dietary behavior, as their target 2.2 postulates to "end all forms of malnutrition by 2030". In Austria, however, one in every five children below the age of 5 is malnourished (overweight).

Relation to climate: From the climate impact perspective, it is uncontested that vegetable products affect the climate to a much lesser extent than animal products, especially meat, do. Globally, the agricultural sector emits roughly 25 percent of all greenhouse gas. The livestock sector alone generates 18 percent of global GHG emissions. In Austria, the agricultural sector gives rise to roughly 9 percent of the country's GHG emissions (excluding the GHG emissions of net meat imports).

Potential: A scientific review covering more than 60 studies concludes that by fundamentally changing dietary patterns, GHG emissions from agricultural production could be reduced by up to 70 percent. Although comparable only to a limited degree with regard to health impacts, the studies showed that the relative risk of a premature death from nutritional diseases can decrease by up to 20 percent. Despite the lack of methodological standards, it can be summarized that a more plant-based diet can help reduce the number of premature deaths and the occurrence of nutritional diseases and curb diet-related GHG emissions considerably (high agreement, robust evidence).

Options for action: Potential action may face resistance from the parties involved for different reasons and despite strong evidence, and the biggest challenge will be overcoming this resistance. The best way to go about it would be to develop a participative and coordinated management of measures in order to avoid effects detrimental to farmers and consumers, for instance.

Scientific analyses suggest that "soft measures" such as information campaigns prove ineffective in changing prevailing nutritional trends. By contrast, clear price signals, accompanied by targeted information campaigns as well as advertising bans, can be highly efficient in achieving changes (high agreement, medium evidence). Price hikes for meat products in the wake of raised compulsory standards in livestock farming, for example, are suited to send out a clear message. The amount of money consumers pay for food will remain constant – with the help of support measures where necessary – as consumers will partially renounce the consumption of expensive meat in favor of lower-priced fruits and vegetables, which will re-balance their budgets to some extent. Similarly, farmers' revenues will remain steady as a decrease in supply will result in a rise in prices per kilo. Alternatively, studies suggest taxes on food based on its greenhouse gas impact, with revenues to be used to compensate for income losses, to support prices of wholesome foods worth promoting from a health perspective, and for the purposes of health promotion in general.

It should be noted in this respect that currently the public is obliged to pay for all social and health care costs driven up by unhealthy diets. The German *Umweltbundesamt* (UBA; Federal Environment Agency) advocates the reduction of VAT rates on fruit and vegetables in order to generate benefits for the climate and public health. The Food and Agricultural Organization (FAO) of the United Nations pleads in favor of taxes and charges to reflect the environmental damage inflicted in order to render livestock production more sustainable. Taxes on animal products in the amount of EUR 60 to 120 per ton of CO_2 implemented in each of the EU-27 may help to save between approx. 7 and 14 percent of GHG emissions from the agricultural sector (high agreement, medium evidence).

An additional approach would be to re-think marking obligations: Rather than putting labels on products that are climate-friendly and healthy, it would make more sense to label products for being harmful to health or the climate.

For starters, an important approach would be to serve healthy and climate-friendly food in state institutions such as schools, kindergartens, military barracks, hospitals and retirement homes, and also in hotels and restaurants (high agreement, medium evidence). To add a further leverage point, it would be advisable to reinforce health and climate literacy in the education and training of cooks, dietitians, nutritionists and purchasers for large food and restaurant chains.

State policies should have a vital interest in climate-friendly and healthy dietary behavior, as it does more than just serve to meet climate objectives because it leads to increases in related labor productivity and to cuts in public health spending, and results in an unburdening of public budgets.

5.2 Mobility

Need for action: Transport is a highly relevant sector in terms of climate and health. It accounts for 29 percent of Austria's GHG emissions, 98 percent of which are caused by road transport (44 percent of the latter from freight transport and 56 percent from passenger transport respectively, 2015). Emissions have risen by 60 percent since 1990 (reference year of the Kyoto protocol) with a disproportionate increase in freight transport. Poor air quality in cities and in alpine valleys and basins remains a problem in Austria, especially with regard to nitrogen dioxide emissions – with this in mind, the EU initiated infringement proceedings against Austria in 2016. Particulate emission and ground-level ozone readings are also exceeding their respective limit values (with recordings of elevated ozone levels at 50 percent of all measuring stations). Emissions are mainly caused by road traffic, and, to a great extent, by diesel-powered vehicles (high agreement, robust evidence). 40 percent of the participants in a survey of the Micro-census conducted by Statistics Austria (Federal Austrian Statistical Office) reported that they were annoyed by noise with road traffic as a decreasing but still dominant source. With regard to freight trains, policymakers have already created incentives by way of noise-differentiated track access charges to have providers put into service quieter brakes (which can help reduce noise by up to 10 dB).

Necessary as it may be, the technological transition from fossil fuel vehicles towards electric vehicles alone will not suffice to meet all the different goals as it fails to redress problems such as the risks of accidents, particulate pollution from tire and brake wear as well as resuspension, noise, traffic jams and land use for road infrastructure. It is especially multi-track vehicles which, due to their high space requirement, hamper quality of life improvements in urban areas – when temperatures are rising in particular. In addition, any major positive impact on the climate footprint will only be felt if power generation for electric vehicles goes climate-neutral. The health potential of climate-sensitive mobility does not stop at electric mobility (high agreement, robust evidence). The SDGs (target 3.6) also require the number of road fatalities to be globally halved by 2020, which cannot be realized through electric mobility alone. Statistics, however, indicate that the number of road fatalities in Austria is on the decline and that, by adopting additional measures, a halving of said numbers seems feasible. A reduction in car use, mileages and driving speed, in particular, is likely to reduce road fatalities and noise pollution, the emission of pollutants and GHGs (high agreement, robust evidence).

Potentials: The shift towards more climate-friendly means of freight and passenger transportation should at any rate become part of the solution. By attracting customers through improved services, it is possible to increase passenger numbers for local public transport and to reduce the modal share of private motorized transport at the same time. Vienna succeeded in reducing its modal share in private motorized transport by 4 percent within less than a decade. A shift towards active mobility (walking, cycling) and public transport will lower pollution and noise emissions and lead to an increase in physical activity, which in turn helps reduce obesity and overweight, minimizes the risk of developing cardiovascular and respiratory diseases and cancer, as well as the risk of sleep or psychic disorders. All these factors taken together lead to an increase in life expectancy with more years of healthy life (high agreement, medium evidence). In addition, the modal shift brings about significant cost savings for public health care. Cost-benefit analyses conducted in Belgium found that the amounts that could be saved in health care were between two and fourteen times higher than what had to be spent on bicycle paths.

A statistical survey of 167 European cities indicates that by extending a bike-path network it is possible to raise the share of bicycle traffic and that, by persistently adapting traffic concepts, it is indeed possible to increase the cycling modal share by more than 20 percent in German (e.g. Münster 38 percent) and Austrian cities (Innsbruck 23 percent and Salzburg 20 percent).

Through scenarios based on tried and tested measures, a study on the cities of Graz, Linz and Vienna indicated that, even excluding electric mobility, deaths per 100,000 population could be reduced by 60 in actual numbers and CO_{2equ} emissions from passenger traffic by 50 percent, whilst curbing annual health care spending by almost EUR 1 million per 100,000 population in the process. This could be achieved through a mix of policies combining the establishment of strolling zones, reduced-traffic zones, the construction of new bike paths and infrastructure, an increase in service frequencies of public transport and cheaper fares in urban/rural transit. Provided power generation goes carbon-neutral and if complemented by electric mobility, all of these measures could save 100 percent of CO_{2equ} emissions and prevent 70 to 80 deaths per 100,000 inhabitants annually (high agreement, robust evidence).

Options for action: Urban mobility in particular offers vast health co-benefits and great potential for improving quality of life, and therefore also is a promising opportunity for climate protection and health improvement. By becoming more people-friendly rather than car-friendly and by embracing active mobility, cities, towns and villages would improve social contacts, well-being and health statuses - and even reduce their crime rates. In addition, the dismantling of roads and parking lots will "depave" the way for creating green areas and for alleviating "heat island" effects in the process. All of these advantages can be taken care of through appropriate measures of settlement structuring, such as physical layouts of homes, workplaces, shopping malls, schools, hospitals or retirement homes, which determines traffic volumes to a large extent, as well as through legal frameworks and guidelines for land use and urban planning (high agreement, robust evidence).

By promoting and incentivizing active mobility and by favoring public transport and sharing models over private motorized transport, it becomes possible to exploit previously untapped potential: active mobility can be promoted by creating low-emission zones of reduced motorized traffic as well as pedestrian promenades and bicycle boulevards, for instance, while, at the same time, parking spaces could be reserved for electric vehicles or approval granted only to car sharing companies committed to electric vehicles. It would eventually be possible to fund and intensify such pull measures through strategies to internalize external costs, in particular those of motorized traffic.

For the mobility sector to be of great benefit to both climate protection and health promotion, the institutional cooperation of all competent municipal, provincial and national authorities is recommended. To be functional, any cooperation first and foremost requires that necessary resources and capacities be made available for the exchange of information and opinions (high agreement, medium evidence).

A lot could also be done with regard to air traffic, which is encouraged and favored by policymakers and which is not covered by the Paris Climate Agreement. Air traffic, undoubtedly, leaves a very large carbon footprint mark and urgently requires action to be taken (high agreement, robust evidence); any plans to reduce it, however, have often been declined due to the various economic interests involved. Air traffic could be reduced by imposing a CO_2 tax on hitherto untaxed jet fuel, for instance, to eventually reduce harmful emissions such as particulate matter, secondary sulfates and nitrates as well as noise and the elevated risk of contracting infectious diseases.

5.3 Housing

Need for action: Housing conditions play an essential role in terms of health, well-being, climate protection and any adaptation to climate change. Both the physical layout (settlement structures) and building techniques trigger long-term path dependencies and have far-reaching implications for mobility and leisure behaviors. Buildings account for roughly 10 percent of Austria's GHG emissions with numbers declining, but the housing stock has been increasing for decades and 87 percent of all buildings are single-family or duplex houses, with apartment buildings containing 3 or more units accounting for a mere 13 percent.

In cities in particular, increasing summer heat loads during the daytime with no significant drops in temperatures overnight lead to uncomfortable indoor air climates and eventual health issues (especially for people in poor health and the elderly as well as children) (high agreement, robust evidence). Other well-documented stress factors include noise and air pollution. Noise levels measured overnight in front of a window which exceed approx. 55 dB(A) can cause health issues such as impaired cardiovascular regulation, mental disorders, reduced cognitive performance or glucose imbalances. Such elevated levels occur regularly on very busy roads (within cities and on highways and expressways) as well as near airports. The option of airing apartments is limited by noise and air pollution as well.

Options for action: To ensure that the focus of urban planning is on health improvement and climate-friendly housing, it would be essential to have climatologists and medical specialists participate in urban planning processes. When it comes to construction and housing policies, climate change adaptation strategies and emissions reduction on the one hand and the traffic situation and green and local recreation areas on the other cannot be addressed separately. While rules and regulations and support measures are increasingly taking account of the effects of climate change, they often fail to acknowledge the close interdependencies between housing and traffic and/or parking spaces.

When it comes to the energy-oriented restoration of old building stock in Austria, rates are as low as 1 percent and the quality of renovation is poor as well. Differences in ownership structures and diverging interests of landlords and tenants are urgent obstacles that need to be overcome. An increase both in the quality (e.g. high-quality thermal insulation, use of comfort ventilation systems) as well as the numbers of premises to be renovated will help reduce heat stress and will produce positive health effects (high agreement, robust evidence), which equally applies to office buildings, hospitals, hotels, schools, etc. This can also help to reduce the need for energy-intensive air conditioning systems. While the call for "affordable housing" may be understandable, it should not, however, translate into "cheap and poor construction design", as poor construction means elevated heating costs, which ultimately affects affordability. In addition, low-emission heating and hot water systems can become key components in terms of climate protection; in densely populated areas, however, they can achieve even more by improving people's health, since they contribute to reducing air pollution (high agreement, robust evidence).

Single-family and duplex houses with their attached garages and road space mean additional sealed surface, materials and energy input, and they imply a long-term commitment to private motorized traffic. As for new buildings, any such constructions should be called into question because of their climate footprint and health implications (high agreement, robust evidence). With a population increase of almost 2 percent and with an increase in sealed areas of 10 percent (approx. 54 acres per day), Austria is at the top of the list in Europe when it comes to soil sealing. As a consequence, rather than aspiring to own a house and garden it would be better to incentivize more attractive solutions such as apartment buildings situated in well-developed, low-traffic areas offering a high quality of life and access to green spaces, which, in addition to the numerous climate and health benefits they entail, help to strengthen feelings of community. What is needed now is the development of suitable passive and energy-plus building standards for larger buildings (high agreement, robust evidence).

5.4 Health Care Sector

Need for action: Accounting for 11 percent of the country's GDP (2016), Austria's public health care system is not only of great importance in economic and political terms as well as for society as a whole, it also has a climate impact; and it is stretched to its financial limits. Ironically, contrary to its true purpose of promoting health, it directly (e.g. through heating/air conditioning and power consumption) and indirectly (mainly through the manufacture of medical products) fuels climate change and its health implications (high agreement, medium evidence). Yet, the question of how to reduce emissions from the public health sector has neither been addressed by Austria's climate and energy strategy nor on an international level. Similarly, the health care reform papers fail to acknowledge its impact on climate change. Although Austria's National Health Targets do refer to sustainably securing the natural resources (Health Target 4), they do not point to the fact that it is essential to reduce emissions from the health care sector. Thus far, some hospitals have implemented efficiency/savings measures to reduce building-related emissions – in part for economic reasons. For the first time, a project within the framework of the Austrian Climate Research Program ACRP is currently gathering data on the Austrian health care sector's share in GHG emissions.

Potential: As for traditional environmental protection, for instance, in building construction, it turns out that GHG emissions are to a large extent caused by intermediate consumption. A carbon footprint study focusing on the US public health sector indicates that 10 percent of the United States' GHG emissions are directly or indirectly attributable to the health care system, with emissions from intermediate consumption exceeding direct on-site emissions, and that the lion's share of GHG emissions stems from the manufacture of pharmaceutical products. Studies in the UK and in Australia paint a similar picture, though indicating slightly lower figures (high agreement, medium evidence).

Avoiding unwarranted or non-evidence-based treatments (and hospitalization) helps not only to reduce emissions (e.g. particulate emissions) from the health care system and their health implications but will also be beneficial in terms of protecting the climate and health in general (high agreement, medium evidence). This includes avoiding medication over- and under-use, multiple or duplicative diagnoses or incorrect assignments (which is when treatment and care do not fit the diagnosis).

Options for action: For health and climate benefits to come to fruition, it is advisable to prepare a specific mitigation (and adaptation) strategy for the health care sector to provide a political guideline for all authorities involved at federal, provincial and organizational levels. And, with reference to the Austrian Health Target 4, any such strategy should aim to reduce direct and indirect GHG emissions, other harmful emissions, waste, and the use of resources as well as to adopt adaptation measures such as the development of climate-related health literacy and to integrate the topics of "climate and health" into education and training programs of health professionals. The various national and international policies can serve as role models during the implementation process [e.g. U.K. National Health Service, *Österreichische Plattform Gesundheitskompetenz* (ÖPGK; Austrian Platform Health Literacy)]. At the same time, participatory structures which allow for an exchange between the various parties involved should be established and form an integral part of the process.

Systematically integrating (if necessary compulsory) eco-quality criteria into quality control and utilizing the incentive mechanisms of the *Gesundheitsqualitätsgesetz* (Health Care Quality Act) can help and support environmental management departments in hospitals in particular.

Significant reductions in GHG emissions, in risks to patient safety and cuts in health care costs could be realized by avoiding unnecessary or non-evidence-based diagnoses and therapies (high agreement, robust evidence). Systematically enforcing the international initiative *"Gemeinsam klug entscheiden"* ("Making Choices Together/Choose wisely") may prove highly promising in curbing medication over, under- and misuse and hold great potential for mitigating economic and ecological impacts (high agreement, low evidence). What is problematic when it comes to avoiding unwarranted diagnoses and therapies is the fact that the pharmaceutical and medical device industries fund medical training programs in Austria with considerable amounts so that programs to a very large extent tend to cater to specific interests.

Consistently prioritizing multi-disciplinary primary care, health promotion and ill-health prevention in accordance with the health care reform can contribute to reducing energy-intensive hospitalizations and, therefore, GHG emissions (high agreement, low evidence). Medical treatments that focus to a greater extent on health promotion can help people adopt healthier diets and increase their physical activities through active mobility – which will also contribute to climate protection. In addition, the handing over of a greater number of patients to regional primary care (registered medical practitioners or health care centers) can reduce GHG emissions, as it reduces the traffic flows of patients and visitors to and from hospitals.

The health care system's impact on the climate will have to be analyzed during these implementation initiatives (e.g. analyses of GHG-intensive medical products and possible alternatives). Given their complex nature, the interdependencies require the conduct of adequately funded research projects on international, interprofessional, inter- and transdisciplinary levels with a focus on practical implications.

6 Transformation at the Intersection between Climate and Health

For the purpose of keeping the health implications of climate change in Austria in check, it will not suffice to enforce technological solutions such as energy efficiency improvements, electric mobility, new therapies or building modernization, nor will such solutions be sufficient to meet the goals set forth in the Paris Climate Agreement, let alone to fulfill Austria's commitment to the SDGs vis-à-vis the world community. Rather, it is necessary to launch a comprehensive transformation process which challenges consumption patterns as well as economic modes of production and our health care system in search for answers in order to set the course for new development strategies that offer appealing qualities of life and equal opportunities for all in accordance with the Sustainable Development Goals (SDGs). Any such comprehensive transformation is bound to face resistance, such as inherent preservation tendencies which often pander to group interests and fail to adequately consider long-term disadvantages and emerging risks for the common good. It is where climate and health issues intersect that new and innovative concepts should be tried out in specific gradual steps of transformation since, in some areas, the health benefits will have a bearing on a great number of people, will set in rather quickly and be accompanied by positive climate effects.

6.1 Initiating a Cross-policy Transformation

Any transformation process will have to be stepwise, reflective and adaptive in nature in order to not run the risk of imposing incoherent individual measures which fizzle out rather ineffectively. All efforts to harness the numerous synergies and to avoid detrimental reciprocal effects will be in vain unless the interdependencies of, amongst other things, heat events, demographic dynamics, traffic, including active mobility, green areas, healthy diets, climate-related health literacy and a more climate-friendly health care system that focuses on disease prevention and health promotion are jointly considered and developed in the process.

It is true that some of Austria's strategies have already taken account of such a transformation process at the intersection between climate and health, but they hitherto failed to gain any satisfactory momentum. At the very least, the following three strategic areas offer synergies to be tapped into: On the one hand, there are the Austrian Health Targets with their aim at high-impact climate benefits (Target 2: Fair and Equal Opportunities in Health, Target 3: Health Literacy, Target 4: Secure Sustainable Natural Resources such as Air, Water and Soil and Healthy Environments, Target 7: Healthy Diet, Target 8: Healthy and Safe Exercise); and on the other, there are the Paris Climate Agreement as well as the recently adopted Austrian climate and energy strategy together with the Austrian adaptation strategy. The climate and energy strategy places the main focus on traffic and buildings, areas highly relevant to public health. Health promotion through active mobility forms an integral part of traffic concepts. And, not least by virtue of ratifying the UN General Assembly's Resolution "Transforming our world: the 2030 Agenda for Sustainable Development" with its 17 development goals and 169 targets, Austria has committed to far-reaching transformative steps in the fields of climate change and public health. The current report by the Federal Chancellery already takes note of the fact that adhering to the Health Targets will contribute to achieving a variety of sustainability goals.

In its latest status report on the environment and health in Europe, WHO/Europe identifies the lack of intersectoral cooperation at all levels as the main obstacle on the road to the successful implementation of climate measures (high agreement, robust evidence). Likewise, the EU calls for the integration of health into climate change adaptation and mitigation strategies of all sectors for the purpose of public health improvement.

Climate policies, in this regard, can become the driving force behind the "Health in All Policies" approach, and health the engine to propel integral transformative steps. What would be needed to realize this potential, however, is a decisive cooperation that can prove successful in Austria based on the aforementioned groundwork that has been laid (Health Targets, climate and energy strategy, Sustainable Development Goals). If rooted in a precise political mandate, climate and health policies could be tied together where topics intersect through the establishment of exchange structures for a transformation process, which, in turn and in addition, can make key contributions towards achieving the sustainability goals in the process. For the implementation process to be rapidly completed it would be necessary to have the government, the provinces and the municipalities participate on a broad basis, and also to include social insurance agencies and academics. Topics that climate and health strategies should address specifically include the complex around heat-buildings-green areas-traffic, a healthy and climate-friendly diet, active mobility, developing health literacy, the emission mitigation and adaptation strategy for the health care system and also the systematic application of the *Umweltverträglichkeitsprüfung* (environmental impact assessment) in tandem with a health impact assessment for regional and urban planning.

6.2 Harnessing the Scientific Potential for Transformation

Even if there is no dispute as to which goals to set for both health and climate strategies – e.g. the lowering of meat consumption, a reduction in air traffic or an urban density increase – no answer has yet been provided as to how the respective measures can be formulated to win over the public and decision-makers and to avoid disadvantages and make the most of opportunities. This requires innovative scientific methods which not only monitor and analyze the systems from the outside but which, by favoring transdisciplinary approaches, help set in motion a targeted participative transformation and launch learning processes that are more likely to result in actual solutions. In any case, it will also be up to the science community to evaluate measures, identify interconnections vital to the success of strategies or to come up with forms of communication suited to accessing groups which are otherwise hard to reach.

To increasingly guarantee that only sensible and assertive action is taken, it is expedient to develop and implement a concept for monitoring the climate impact on all ecospheres and on health. To further the understanding of the direct and indirect effects of climate change, it stands to reason to establish and operate test regions. And it is also advisable to draw up a comprehensive population register following the example of Scandinavian countries in order to get a firmer grasp on climate vulnerability and the harmful effects of climate change thus far.

In addition, and to further ensure that appropriate action is taken, it will also be necessary to fill the gaps in our knowledge of where the issues of climate change, demographics and health intersect. This includes the collection of data on the emissions from health services (including intermediate consumption), the preparation of mitigation measures, and lifecycle assessments of medical products, in particular drugs, to help evaluate side effects such as the climate impact of medical treatments in relation to the treatment outcome (whether any success of a treatment is worth the damage). There is also a need for analyses of the efficiency of monitoring and early-warning mechanisms that focus on the reduction of harmful effects; in this context, however, some methodological questions of how to quantify treatment success have yet to be answered (e.g. measurability of a reduction of mental trauma incidents).

Both medical as well as agricultural research would do well to improve transparency with regard to their scientific problem-solving techniques, test and trial designs, and also sources of funding, since research and education in both disciplines are increasingly funded by interest groups and the economy. This could be a necessary step towards an actual reduction in overmedication and multiple or duplicative diagnoses.

With buildings becoming increasingly hi-tech in order to improve energy efficiency, it will be necessary to analyze whether this will cause new health issues and how much net GHG emissions are actually saved as soon as intermediate consumption is factored into the carbon footprint.

With demand for quality food at lively levels, organic farming may help to achieve the Paris Climate Agreement objectives. An assessment, however, would require conclusive scientific data on the health effects of organic vis-à-vis non-organic food.

Ultimately, there is much to be learned from initiatives which have already adopted healthy and climate-friendly lifestyles. Such initiatives include, for instance, ecovillages, slow food or slow city movements and transition towns. Attractive and suitably convenient lifestyles can be incentivized by reducing negative and enforcing positive factors. Multi-faceted transformation research as well as research-oriented teaching could accelerate the respective transformative developments and could, thus, reduce impeding and promote beneficial factors to foster acceptable and viable developments that improve the quality of life.